The Field Journal for Cultural Anthropology

Jessica Bodoh-Creed

California State University, Los Angeles

Los Angeles | London | New Delhi
Singapore | Washington DC | Melbourne

FOR INFORMATION:

SAGE Publications, Inc.
2455 Teller Road
Thousand Oaks, California 91320
E-mail: order@sagepub.com

SAGE Publications Ltd.
1 Oliver's Yard
55 City Road
London EC1Y 1SP
United Kingdom

SAGE Publications India Pvt. Ltd.
B 1/I 1 Mohan Cooperative Industrial Area
Mathura Road, New Delhi 110 044
India

SAGE Publications Asia-Pacific Pte. Ltd.
18 Cross Street #10-10/11/12
China Square Central
Singapore 048423

Printed in the United States of America

Library of Congress Cataloging-in-Publication Data

ISBN: 9781544334110

This book is printed on acid-free paper.

MIX
Paper from responsible sources
FSC
www.fsc.org
FSC® C008955

Acquisitions Editor: Joshua Perigo
Editorial Assistant: Noelle Cumberbatch
Production Editor: Astha Jaiswal
Copy Editor: Lana Arndt
Typesetter: Hurix Digital
Proofreader: Ellen Brink
Cover Designer: Gail Buschman
Marketing Manager: Kara Kindstrom

19 20 21 22 23 10 9 8 7 6 5 4 3 2 1

CONTENTS

PREFACE

Introduction to Cultural Anthropology is one of my favorite classes to teach. It allows me to show students why I fell in love with the discipline. Anthropology, at its heart, is about discovering a new culture, embracing differences, and learning to see things that may be familiar anew. Because anthropology begins with Franz Boas falling in love with Inuit culture and deciding to forget what he was originally there to study, anthropologists and those drawn to study anthropology are intrinsically curious observers of the world around them. They want to know why people do the things that they do, especially the things that to the anthropologist seem counter-intuitive. All anthropologists enjoy traveling and discussing the cultures they specialize in studying. Often, they love to learn languages, and they appreciate being let into someone else's world to share in their food, their jokes, their lives, and their culture.

Most importantly, we, as a discipline, refrain from judging others, something that happens far too often in our culture today. Anthropologists ask *how* and *why* questions of the people they study to better understand their practices. They do not judge and malign them or call them "weird" for the things that they do. Just because something is not what *you* would do does not make it wrong. Hopefully the activities in this fieldwork journal make you think about the world around you in new and different ways. I hope they make you ask questions of yourself, push you out of your comfort zone, and get you thinking deeper about your culture, traditions, and practices, and make the familiar, strange and the strange, familiar.

EXERCISE GUIDE

Within each of the exercises in this book, you will find *instructions and notes* on how to assign activities, and *key terms*. Please reference the table below if you have questions regarding what these features mean, or how they can help you navigate through the exercises.

Feature	Purpose
Key Terms Boxes	Please be aware of the anthropological definitions of these terms. Write out the definitions or make notes on what they mean, because you will be asked to apply them in your responses to the questions. The best place to find these terms is to search the glossary of your intro to cultural anthropology book, or there's this thing called Google. Wiki is not a good idea.
Note Boxes	These are notes that will clarify for the instructor and the student how the exercises can be assigned, applied, and submitted.
Read:	Pretty straightforward. Read the material because you will not be able to do the assignment without it.
Field Work Activity	These are assignments designed to get you out of the classroom and into the world. You cannot experience anthropology without seeing, listening, smelling, tasting, and feeling the world around you, even if it's just looking at the world right outside your door.
Field Notes Boxes	Taking field notes, really good field notes, is essential to anyone doing research, especially anthropologists. Use these as opportunities to answer the questions in the exercises, reflect on what you observe, and learn—whether it be out in the world or even within your classroom from other students. The more detail, the better when it comes to field notes.

ACKNOWLEDGMENTS

I would like to thank Rich Warms for his support and guidance. His feedback and excellent textbook were a guidepost through this project. I enjoyed greatly the collaborative spirit with which this project began.

I would be remiss if I didn't mention that I TAed a *lot* of Cultural Anthropology classes in my PhD. I would like to thank Juliet McMullin, Paul Ryer, Sally Ness, and Derick Fay, whose classrooms, style, and character I was able to observe and then model in my own classroom. My colleagues and cohort from UCR were also sounding boards, collaborators, and friends who helped to shape me and one or two of these exercises that have been used more than a couple of times since. So thank you to Theresa Barket, Jon Spenard, Alicia Bolton, Jay Dubois, Shyhwei Yang, Phuoc Duong, Priscilla Loforte, Michelle Butler, and Celia Tuchman-Rosta.

I also was a student in my first anthropology classes taught by Richard Reed, Jennifer Mathews, John Donohue, and Michael Kearl at Trinity. Drs. Donohue and Kearl have since passed on and their legacy is always felt by their students and in projects like this one.

I would like to thank my editor at SAGE, Josh Perigo. You have been an amazing support through this process, and I couldn't wish for a better editor and collaborator.

I would like to thank the reviewers,

Carol Hayman, Austin Community College – Northridge Campus

Annie Melzer, Northern Kentucky University (fka Northern Kentucky State College)

Anne Schiller, George Mason University (fka George Mason College)

Catherine Fuentes, University of North Carolina at Charlotte

Andrew Workinger, University of Tennessee

Their time and consideration were essential to this project.

I cannot miss thanking my husband Corey and our daughter. Their love and support make my world go round.

ABOUT THE AUTHOR

Jessica Bodoh-Creed is a lecturer in anthropology at California State University, Los Angeles. Her research focuses on knowledge production and medical media, specifically the Patient 2.0 or Smart Patient movement. Professor Bodoh-Creed's research looks at four lines of media (fictional medical television, celebrity physicians, cyberchondria and health web searches, and pharmaceutical advertising) to better understand the history of medical media and how this medical knowledge is produced for Smart Patients.

Professor Bodoh-Creed has done ethnographic research in Honduras, one archaeology dig in Mexico (it wasn't for her), worked with primates such as baboons and chimpanzees (also not for her), and finally has worked in media and urban anthropology. Her newest research with her students is on gentrification in Los Angeles. She loves teaching and encouraging her students to go out and be good people in the world. She believes very strongly that the empathy learned by studying anthropology could go a long way toward making the world a better place.

A native Texan, Professor Bodoh-Creed earned her B.A. in anthropology from Trinity University in San Antonio, Texas, her M.A. in sociocultural anthropology from California State University, Los Angeles, and her PhD. in anthropology from the University of California, Riverside. She really likes anthropology. She lives in Los Angeles with her husband, daughter, and their dog.

INTRODUCTION

Anthropologists study culture, and fundamentally, they look at people and their habits and practices. One of the tasks of this fieldwork journal will be to ask you, the student, to look at yourself and the world around you in new and different ways in the hopes that you begin to see things anthropologically. As you work through the exercises, keep in mind that you are continually being asked to work collaboratively with your peers, people who will be coming to the classroom with their own differing cultural backgrounds and experiences. While there may be points where the class sees things very similarly through an American culture lens, there will also be moments where you or one of your classmates might have a very different perspective based on a belief, an experience, or a cultural identity. As anthropology students, you will need to keep in mind that anthropologists refrain from the judgment of others. We seek to understand other cultures and to know why people do what they do, no matter how outside of your own comfort zone or worldly experience it might be. With this in mind, think about your own culture as you embark upon this course. Keep in mind that everyone has culture. It might be defined as something as simple as American culture, or it might be a triple hyphenate. It could be regional or religious, based in sexuality or gender, or historical from your parents. Culture is complicated, in a really good way.

CLASSROOM ACTIVITY: WHO ARE YOU, ANTHROPOLOGICALLY?

©iStockphoto.com/borchee

Note: This exercise can be administered using clicker questions or a survey during classroom lecture with raised hands. It may also be administered by assigning students to check each box that applies. Signal (clicker or hand) if you consider each question to be true.

(Please check all that apply)

1. I consider myself to be an American. ❒
2. I have a cultural identity other than "American." ❒
3. I have a second or third identity that connects to my American culture as well. ❒
4. I have family members who live outside the United States and do not consider themselves Americans. ❒

5. I am a legal citizen of more than one nation. ❒
6. I have a well-defined idea of my “culture.” ❒
7. I have membership to several cultures. ❒
8. I have wondered about my own culture and family history. ❒
9. I do not feel connected to any one single culture. ❒

Field Notes

Discuss in small groups or reflect on your own how and when we feel membership to a culture, and what might connect someone to a given culture or cultures.

INDIVIDUAL ACTIVITY: WHO ARE YOU, ANTHROPOLOGICALLY?

©iStockphoto.com/borchee

Question 1: What is your culture? Please list the culture or cultures you identify as being part of.

____

____

____

____

____

Question 2: How do you define your own culture(s)?

____

____

____

____

Question 3: How do you know this is your culture? Who helped shape your belonging to this culture or these cultures?

EXERCISE ONE

CULTURAL RELATIVISM AND ETHNOCENTRISM

Key Terms

Cultural relativism: ______________________________

____

____

Ethnocentrism: ______________________________

____

____

Emic: ______________________________

____

____

Etic: ______________________________

____

____

____

CLASSROOM ACTIVITY: WHAT WOULD AN ANTHROPOLOGIST DO?

Note: This is designed for a classroom discussion in a small class or for group discussion in a larger class.

Read: Because most anthropologists must travel for their research, we end up in some pretty different, fun, or even scary situations. Some of us may research dangerous topics like wars, violence, and natural disasters. However, anthropologists with less physically challenging research questions also face danger from the environment, disease, politics, climate, or many other things. How we handle these situations, no matter how closely they affect us, can have repercussion on our fieldwork and informants. We are personally involved in the lives of those who teach us about their cultures, and they inevitably become involved in ours, many becoming lifelong friends.

Activity: Do this thinking exercise ahead of class.

Imagine you are an anthropologist doing field work in a remote jungle in northeast Brazil, hours away from Western medicine and without a way to contact anyone for help, and you find yourself very ill. You aren't exactly sure what you might have contracted. It might be malaria, it could be typhoid, or maybe it is some sort of parasite or amoeba. A local who has been your guide and helpful friend offers to take you to a traditional healer who can diagnose you and give you local herbs to make you feel better. This is what everyone here does, and your friend highly recommends it. You feel just terrible, so you go the local healer with your friend. The healer looks at you, takes part of a rattlesnake, some skin and bones from a fox, and cow urine, and combines them with herbs you don't recognize into a mixture and tells you that you must drink it over the course of 3 days.

Question 1: What do you do? Do you drink it? Do you trust the potion? Why or why not?

___

___

___

___

___

___

___

Question 2: Think about how your new trusted local friend will feel if you hesitate or seem to judge the potion or the healer. Would that affect your ability to do research in the village?

Question 3: Now imagine you are in Nigeria in West Africa in the same situation, and the healer offers you a concoction made of pangolin, one of the world's most trafficked and endangered animals. Would you take the potion?

©iStockphoto.com/andyschar

©iStockphoto.com/2630ben

Question 4: How do these two situations bring up the idea of cultural relativism?

Question 5: If you declined the pangolin treatment, would you be judging Nigerian cultural use of animals in traditional healing?

____

____

____

____

____

Question 6: Are you using an emic or etic perspective to think about this? Please explain.

____

____

____

____

____

____

____

Discuss: Now discuss your answers in small groups. What do your classmates think about this? Are you all in agreement?

Field Notes

Read: When they think about cultural relativism, anthropologists try to consciously consider why cultures engage in particular practices. Humans are generally logical in their thinking, and often habits, tendencies, and rules are the result of context and historical circumstances. So when anthropologists consider exotic practices like cannibalism, which has been discussed, critiqued, and sometimes held up as an example of the craziest or least logical human practices, anthropologists have to be especially careful. We need to understand and think through the few real examples we have of these practices. In 1979, anthropologist William Arens, published a book called *The Man-Eating Myth* in which he claimed there is no real evidence of cannibalism, except in rumors about others. Arens argued that accusations of cannibalism were ways that people denigrated others. He noted that they were often used by Europeans to justify slavery and colonialism.

Arens' argument proved to be a case of "just because you're paranoid doesn't mean they're not out to get you." In other words, Arens' argument was incorrect, although it is true that people have sometimes been incorrectly accused of cannibalism to denigrate them and reinforce the case for colonialism and slavery. It is also true that research in cultural anthropology, medical anthropology, and archaeology have established beyond reasonable doubt that numerous human societies have practiced cannibalism. There are many reasons people practice cannibalism but, in many cases, it is performed as part of funerary rituals to honor group members.

Of course, there have been some well reported episodes of cannibalism in American history and recent world history as well. The Donner Party on the Oregon Trail turned to cannibalism when it was stuck in the high Sierra Nevada mountains in the winter of 1846–1847. Far more recently, when Uruguayan Air Force Flight 571 crashed in the Andes in 1972, 16 passengers and crew members survived by cannibalizing their dead.

INDIVIDUAL ACTIVITY: YOU ARE WHAT YOU EAT

Note: This can be a reflection assignment that students write and turn in or a forum quiz/post in their Learning Management System.

Read: Imagine you begin working with a group of people in another country, on another continent, and they confide in you that they eat their dead. Cannibalizing the dead helps them pass on to the afterlife, and they want you to join their practice after a friend of yours within the group dies.

Question 1: Would you join in? If so, why, and if not, why not?

____

____

____

____

____

Question 2: Assuming you have chosen not to engage in cannibalism, how do you stay culturally relative and respect people's customs while seemingly disapproving of their practice of killing and eating other humans?

____

____

____

____

____

Question 3: What do you think an anthropologist should do? How does one refrain from ethnocentrism in this kind of situation?

____

____

____

____

____

Self-reflect: Remember, anthropologists may have to consider practices like cannibalism, and in these cases, they have to be especially careful.

Under what circumstances would you eat human flesh? (Circle each one that applies)

- Only when in a dire situation where I would die if I did not eat food and that is my only option.
- If I am in another country and culture, and this is their practice, and it would be rude not to do it.
- Any and all circumstances. That could be really awesome to experience.
- Only if it is leg meat. Yum.
- Never. That's too weird.
- OMG, who does that? Yuck.

EXERCISE TWO

ETHICS AND ETHNOGRAPHY

Key Terms

Informant: ______________________________

Institutional review board (IRB): ______________________________

Fieldwork: ______________________________

____

Read: Anthropologists study all kinds of things, from strawberry farmers in California, to robots that fight, to video game designers, to remote villages in Africa. It is hard to find things that anthropologists haven't studied. Anthropologist Kim Jenkins even studies cargo shorts and why American men love them. Really. Anthropologists who are also college professors often study their students. University students are fascinating creatures, and what it means to be a student now is very different from what it was 10 years ago, 20 years ago, and 50 years ago (when some of your professors were last in college, ha!).

Who anthropologists are as people affects how they are perceived and regarded in the field. This is something Renato Rosaldo calls positioning. If you are a female anthropologist, your gender may raise an issue with the culture you want to study, and so your positioning could change your ability to study certain things. For example, a male anthropologist would have a hard time perhaps asking adolescent girls about their first menstruation, or a female anthropologist might have difficulty asking adolescent males about their first attempts at flirtation with girls. Who we are can affect our ability to study certain topics, to gain access to some places, and to get people to trust us and open up about private topics.

CLASSROOM ACTIVITY: STUDYING STUDENTS

©iStockphoto.com/skynesher

Note: This is designed for a classroom discussion in a small class or for group discussion in a larger class. This exercise can be administered using clicker questions or a survey during classroom lecture with raised hands. It may also be administered by assigning students to check each box that applies.

Signal (clicker or hand) if you consider each question to be true.

Signal (clicker or hand) if you consider each question to be an issue.

Read: Think about this scenario: An anthropologist wants to look at drug use and sexual activity on a college campus. He or she will be conducting focus groups and interviews with college students about their extracurricular activities as they relate to drugs and sex. The researcher plans to ask about intimate details and histories of the informants. Which of these details about the researcher will become important as to whether anyone will feel comfortable talking to them in an open and honest way?

(Please check all that apply)

1. The researcher is the same age as the interviewee. ❒
2. The researcher is older than the interviewee. ❒
3. The researcher is much older than the interviewee. ❒
4. The researcher is the same gender as the interviewee. ❒
5. The researcher is a different gender than the interviewee. ❒
6. The researcher is the same race as the interviewee. ❒
7. The researcher is a different race than the interviewee. ❒
8. The researcher seems kind. ❒
9. The researcher seems serious. ❒
10. The researcher seems trusting. ❒
11. The researcher seems to judge the answers people give. ❒
12. The researcher talks about their own drug use. ❒
13. The researcher talks about their own sexual activity. ❒
14. The researcher seems to care about the respondents' answers. ❒
15. The researcher seems not to care about the respondents' answers. ❒

Field Notes

1. Why do these traits influence your willingness to provide information?

2. What matters most to collecting honest, real information from participants?

3. What kind of person and personality might make a difference in what you would share with the researcher?

INDIVIDUAL ACTIVITY: RISKY RESEARCH

©iStockphoto.com/joshuaraineyphotography

> **Note:** This can be a reflection assignment that students write and turn in or a forum quiz/post in their Learning Management System.

Read: Institutional review boards (IRBs) are an important part of the research process for all social and natural scientists who work with human subjects, and anthropologists are no exception. IRBs are found in universities, research institutions, hospitals, pharmaceutical corporations, and government agencies, all organizations that work with people in social research or medical testing settings. IRBs are committees of people who look at prospective studies by faculty, researchers, agencies, doctors, nutritionists, pharmaceutical researchers, biomedical researchers, and others and decide whether the study is based on sound science, fills a need within existing research, has a grounding in medicine or social testing, and also—which is very important—whether it places the participants and researchers at risk. The risk that participants face has to be reasonable, and the benefits have to outweigh the risks. Institutional review boards are guided by both government regulations and their own rules. Some are more conservative and others more flexible, but all are concerned with assessing risks to participants and the benefits of research. IRBs were created after several early 1900s medical researchers conducted studies that were based on racist, sexist, homophobic, and other similarly prejudiced views and in turn hurt people, acted criminally, and even caused participants' deaths. For more information on this, look at Nazi human studies, the Tuskegee experiment, and literature on forced abortions in the United States' prison system. IRBs now protect populations from harm and make sure the studies are strongly grounded in established science. Anthropologists work diligently to craft IRB proposals that demonstrate that they will limit the exposure to risks for their subjects and ensure that the benefits of the study outweigh the risks to participants.

Question 1: Imagine you are an anthropologist designing a study on a homeless population near campus. What kinds of ethical considerations must you think about when preparing to do your ethnographic project? Write down two things that you should do so you can maintain a good relationship with this community and protect them and their privacy while also getting information about their lives and their culture.

Question 2: How would who you are as a person (your demographic information, personality, approach to the study) change the outcome of the research or possibly give you more or less access to the informants?

EXERCISE THREE

CULTURE

Key Terms

Culture: ________________________________

Norm: ________________________________

Symbol: ________________________________

____

Read: There are six agreed-upon characteristics of culture. These include things like culture is learned and shared among members. One of the most interesting and often delicious parts of field work is eating the local culture—I mean cuisine. I have even known anthropologists to select field sites based on what they would want to eat for the year or more that they will be in the field doing research. Food is so often a reflection of cultural practices, local ingredients and tools and materials available, and it exemplifies gender relations and norms. Have you given a lot of through to what you eat? Most people eat the same approximately 30 things in their weekly food habits. Do you eat the same thing for breakfast every day? Maybe you eat out a lot for dinner and change up the cuisines you eat. Is there a type of food that you would like to try but have not yet because it intimidates you in some way (maybe the menus are all in another language, or there are risky ingredients or you can't control the level of spice or heat)?

CLASSROOM ACTIVITY: A TASTE OF CULTURE

©iStockphoto.com/courtneyk

Note: This can be assigned as a classroom discussion in a small class or in groups in a larger class. It may also be administered by assigning students to complete and reflect.

FIELD WORK ACTIVITY

Read: For homework go out to eat a cuisine that you have never tried before. Think outside the box and look for something that is way outside your comfort zone. If you are an adventurous eater, find a restaurant you have not yet tried and a dish or exotic ingredient that has not yet graced your plate. While there, think about the culture of the cuisine and how it might be reflected in the décor, language, ingredients, and preparation of the food. Feel free to ask your server or the chef questions about the meal and the culture surrounding your meal. Then answer these questions.

Field Notes

Question 1: Where did you go, and why that restaurant?

____

____

____

Question 2: Whom did you bring with you?

____

____

____

Question 3: Was the meal culturally different from your own style of meal preparation, utensils, ingredients, traditions, and flavors? How so?

Question 4: How was the experience? Did it take you outside your comfort zone?

Note: Students may discuss their experiences together in class. Have students get together in groups based on the cuisine they tried to talk about their dishes and experiences.

INDIVIDUAL ACTIVITY: WHERE ARE YOU FROM?

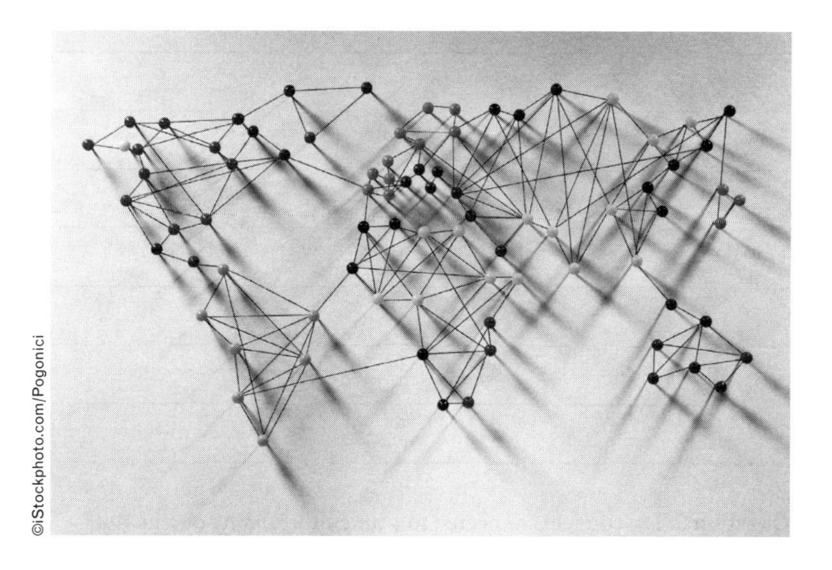

Note: This can be a reflection assignment that students write and turn in or a forum quiz/post in their Learning Management System.

Read: In American society, "Where are you from?" is often a loaded question. It may mean simply what it says, and it is asking for a geographic location that you have lived. It may be that the person is really asking about the ethnicity, race, or home country of someone but does not know how to ask that. However, oftentimes it carries the implication that the person being asked is "not from here." It can be a way of denying identity, asserting difference, and giving the message that someone does not belong. This is particularly the case when the person being asked does not look like or speak like the questioner. We often flounder in trying to ask people about their cultural heritage. Our connections to culture and family history can be very present or feel often distant and remain connected to older generations. Thinking about cultural history, generations in families often identify differently among them because of differences in personal experience, frequency of visits to a home country (or lack thereof), language fluency differences, and many other reasons.

Question 1: Define your culture. What cultures are you connected to? What history and heritage information do you have for yourself?

Question 2: Do you feel connected to your culture? Why or why not?

Question 3: Do your parents or grandparents feel more or less connected to the culture(s) you identify with?

Question 4: Have you ever had the "where are you from" question and felt that it was used to make you feel different, uncomfortable, or like an outsider? What about that question or situation made you feel that way?

EXERCISE FOUR

LANGUAGE

Key Terms

Sapir–Whorf hypothesis: ______________________________

____

____

____

Dialect: ______________________________

____

____

____

Code switching: ______________________________

____

____

____

Read: The Sapir–Whorf hypothesis says that language is a toolkit for understanding culture. For Edward Sapir and Benjamin Whorf, if you know how to speak the language of the people you are studying, then you can better understand the culture of the people. This is one of the reasons why anthropologists must study foreign languages in their formal training and demonstrate proficiency before leaving to do fieldwork. Only by knowing exactly the words, intention, and phrasing can you truly report back with authenticity what informants confide or explain to you.

It is also extremely important in language to know and understand what is important in a culture based on what words people use. People create words for things that they need and for aspects of their environment that are meaningful to them. If you don't need it or care about it, you don't talk about it. Think about cultures like the Gebusi in Papua New Guinea where before globalization they had five words for measuring the amount of things: *zero*, *one*, *two*, *two plus one*, and *many*. What does that tell us about the importance they placed on stuff and ownership? For the Gebusi, talking about any more than two of something was just bragging, and it was probably going to rot in the jungle heat and humidity anyway. There are many reasons why people may not have a word for something, but often it is either because the idea behind the word is not important to them, or it refers to something that does not exist or is not seen commonly in that environment or culture.

CLASSROOM ACTIVITY: HOW CAN YOU SAY THAT?

©iStockphoto.com/Cecilie_Arcurs

Note: This can be assigned as a classroom discussion in a small class or in groups in a larger class. It may also be administered by assigning students to complete and reflect.

Read: Think about how you would talk about something that seems commonplace to you if the words for it did not exist.

Question 1: How would you talk about two of something if the word *two* did not exist?

Question 2: How would you talk about being hot if the word *hot* did not exist?

Question 3: How would you talk about the color blue if the word *blue* did not exist?

Question 4: What are new words that American culture has invented recently?

Question 5: Why these words? What holes in our language did they fill?

INDIVIDUAL ACTIVITY: SAY WHAT YOU MEAN

©iStockphoto.com/PeopleImages

> **Note:** This can be a reflection assignment students write and turn in or a forum quiz/post in their Learing Management System.

Read: Code switching is the ability of speakers to move seamlessly between two languages. Those who code switch use each language in the setting that is appropriate to it. One of the more common ways that we speak about code switching is when we reference someone speaking Spanglish, meaning they speak both Spanish and English and move between the two easily. Almost all of us have some experience of code switching: Either we do it ourselves or we observe others doing it. Code switching is often used when a word in one language does not exist in the other, so you drop that particular word—in, say, Arabic—into an English sentence. Or I have found that cursing in a foreign language to almost always be the preferred word when someone speaks a second language, no matter the language. Sometimes code switching is used to leave people out of a conversation, but be careful with this tactic as that means you are prejudging someone's knowledge of language based on the way they look. More times than I can count, students have stories of someone switching languages to leave them out, only to jump in and tell them in the second language that

that they do not appreciate this. Consider the way you use language in general and your experience of code switching in particular as you answer the following questions:

Question 1: What language(s) do you speak?

____

____

Question 2: In hierarchical societies, the most powerful group generally determines what is "proper" in language. Indeed, the grammatical constructions used by the social elites are considered language, whereas deviations from them are often called dialects. Do you speak any dialects?

____

____

Question 3: Do you utilize code switching in your everyday conversations?

____

____

Question 4: When exactly do you code switch languages? Do you code switch in a single speech event, maybe for specific words? Do you code switch between speech events? Describe when and why you code switch.

____

____

____

____

____

____

Question 5: Have you ever been around people who code switch in front of you? How did it make you feel?

Question 6: If you speak only one language, do you wish to speak another language as well? Why?

Question 7: What benefits would it give you to know more than one language?

EXERCISE FIVE

ENVIRONMENT

Key Terms

Subsistence strategy: ______________________________

____

Agriculture: ______________________________

____

Horticulture: ______________________________

____

Pastoralism: ______________________________

____

Industrialism: ______________________________

____

Foraging: ______________________________

____

CLASSROOM ACTIVITY: FINDING OUR FOOD

Note: This can be assigned as a classroom discussion in a small class or in groups in a larger class. It may also be administered by assigning students to complete and reflect.

Read: Subsistence strategies are important modes for securing nutrition and reliable food sources. There are many different ways that people get their food around the globe. Some will grow small family gardens, others will walk long distances to keep their herd of animals fed, and some will hunt and gather their meals. Horticulture describes those who keep small family gardens, and while we sometimes have gardens, they are often not enough to keep us in food year-round. Horticulturalists, like the Kayapó of Brazil, rotate crops, fish, and trade with neighbors. Pastoralists are those who rely on herd animals like cows, goats, and sheep for subsistence. Cows are difficult to keep because they have long gestation periods and don't make enough of themselves to eat them regularly. Pastoralists who keep cows, like the Nuer of Sudan, often rely on their milk for food, their dung for building houses and their skin for clothing and eat their meat only when it is a special occasion. Modern hunter gatherers, like the Hadza of Tanzania, hunt large game and also birds with bow and arrows and gather berries and roots. It is a hard life, dependent on skill and training to make ends meet.

In the United States, only a very small percentage of the population is involved in food production and usually that involves large scale, commercial agriculture food production. This means that farmers do not eat what they grow, as it is all for sale. Most Americans work a job that pays us money, and then we pay for food that someone else has farmed or grown or raised and slaughtered, often in places far away from us. This is referred to as capitalism, powered by industrialism. Machines speed up modern life and free people to work jobs in sectors like service industries where no one actually makes anything at all. We often are very distanced from the production of our own food, and our only interaction with it is in a grocery store or maybe a farmers market.

Activity: Where does your food come from?

Get into groups of four to five people and research online one fresh food item and one packaged food item that you regularly consume. Examples of these might be avocados and canned tuna fish or romaine lettuce and cashew nuts. Look up where these raw foods are grown, produced, and farmed. What kinds of conditions do the crops need, and what kinds of conditions exist for the workers who have to pick, harvest, and prepare these foods? Who is involved in the production of your food, from the corporations down to the field workers?

Write a paragraph about what you find and present it to the class or post the paragraph along with some of the links you used in your research to a Learning Management Strategy (LMS) forum for this activity.

Field Notes

INDIVIDUAL ACTIVITY: IT'S A FARMER'S LIFE FOR ME

©iStockphoto.com/NNehring

Note: This can be a reflection assignment students write and turn in or a forum quiz/post in their LMS.

Read: While we are mainly capitalists who work in a system of industrialism, you may have grown up in a family that had a small herb and vegetable garden. Maybe some of you were raised on a farm or have been around ranches and livestock. A lot of you probably have not given any thought to what that life might be like to raise animals or grow crops for a living. Farm life is often a hard life dependent on weather, seasonality, the ability to find cheap labor, and the success or failure of your competitors to set pricing. Did you know it is legal in four states (North Carolina, Tennessee, Kentucky, and Virginia) for children as young as 14 to work full time in tobacco fields? Those children test positive for nicotine levels as high as a pack a day smoker. Factory farming of animals is on the increase in the United States as we want cheaper meat in our grocery stores. This puts pressure on producers to make animal production as cheap as possible.

Thus, factory farms often employ women, children, and workers who are here illegally and can be exploited and paid less than the legal minimum wage. These kinds of conditions make this work often back breaking, low paying, and unsafe with the use of pesticides and heavy equipment and other risks from working outdoors in the heat. It is often not remotely like the romanticized, hipster, urban sustainable farm or winery spoken about wistfully in modern culture.

Question 1: You are in college, working to have a career and a good life for you, your family, and a family you might build one day. Would you ever switch subsistence strategies?

____

____

____

____

____

Question 2: Would you work in a tobacco field or on a chicken farm? Why is it something you would or would not want to do?

____

____

____

____

____

Questions 3: Think about how different workers' lives might be from yours in terms of expectations of income and earning potential and the environment in which they and you would work.

What kind of life do you think these workers have? What are they paid for this work?

____

____

____

____

____

EXERCISE SIX

ECONOMICS

Key Terms

Balanced reciprocity: ______________________________

____

____

Reciprocity: ______________________________

____

____

Economic system: ______________________________

____

____

Redistribution: ______________________________

____

____

CLASSROOM ACTIVITY: WHEN JOBS ARE ON THE MOVE

©iStockphoto.com/Liuser

Note: This can be assigned as a classroom discussion in a small class or in groups in a larger class. It may also be administered by assigning students to complete and reflect.

Read: Since the 1980s, globalization has meant that an increasing number of factory jobs have moved abroad, mainly to Asia and Central America. Many of these jobs are found in export processing zones in relatively poor nations, places that have special, favorable tax and tariff rates. Most export processing zones are close to national boundaries, major seaports, or sometimes international airports. The employees who make the products are paid less than similar workers in wealthy countries and often have less job security, work under more dangerous conditions, have few benefits, and no chance at unionization. Employees in these factories often work long days, usually at least 12-hour shifts, if not longer. In America, workers are protected by laws governing workers' rights and the Occupational Safety and Health Administration, an agency that protects workers and oversees dangerous work sites. In many poor nations, protections like that simply do not exist.

Factories in Mexico that build things for import to the United States are called maquiladoras, and they line the border between the United States and Mexico, from California to Texas. They congregate alongside big cities to use their populations as workers and their infrastructures and to be able to ship goods to the United States by truck or train. Workers in Mexican maquiladoras are paid approximately $5 a *day*. Minimum wage in the United States is $7.25 per *hour*. The maquiladora factories build everything from cars to toys to TVs and seasonal items. Recently, as the United States has cracked down on fast-fashion companies like Forever 21, who used to use illegal garment factories in places like Los Angeles, the industry has moved some garment production to Mexico.

Activity: Instructions for a factory closing in the United States and moving to Mexico

You will all be assigned one of the roles listed below. Follow your instructions and simulate a plant closing and picking up shop to Mexico. Work together to see how this process would play out in real life and affect everyone involved. Be creative. Think about what all you have learned thus far in the class. Take notes on your group so you can report the findings to the class at the end. Think about how the company moving from America to Mexico affects families of workers and the culture of each group.

Prompt: You are all part of a production cycle. You will be split into two groups. One from a jeans factory closing in the United States, and another for the jeans factory the company is about to open in Mexico. What happened to these people? What were their lives like? What will they do now?

Characters

American bosses

American workers and their families

Anthropologists

Mexican bosses

Mexican workers and their families

Note: The assigned characters for your class are gender specific, as you will see below. Please consider the conditions and reality that factory workers in these situations have to deal with.

American Bosses (2 male students)

You are in charge of the US factory. You make a good salary of $60,000 a year. You like your workers, but you can see that the company is not making enough money and will have to move production to Mexico. How do you tell your employees they have been laid off and what happens to them now? You keep your job and will oversee 4 people staying to make custom jeans to order. You need only the best workers. Select your new employees from the group.

American Workers and Their Families (1/2 the class)

You have been working at the factory for more than 20 years. You make just enough money to keep your family afloat. You make about $25,000 a year. Your boss lays you off, and there are no other jobs available in the area in manufacturing anymore. What do you do? How will your family react? You can compete for the 4 positions left at your factory that will now pay $40,000 a year making custom-to-order jeans. What will make you competitive? How do you get one of these positions?

Mexican Bosses (2 male students)

You have just been hired to be the manager of a new factory with an all-female workforce. You make a good living of 250,000 pesos ($25,000) a year. What kind of issues will you face with your job? What are you looking for in your workers (e.g., skills)? Interview them for these jobs.

Mexican Workers and Their Families (1/2 the class)

Congratulations! There is a new factory opening up 3 hours away from your home. It is hiring a lot of people. The salary it is offering is 30,000 pesos ($3,000) a year. What kinds of issues will you and your family face when you get a job there? What do you need to do to get the job?

Anthropologists (depending on class size, 2–5 students)

Your job is to find out what is happening to each of the groups above and find out what is going on with everyone since the plant closed. Report back with your findings.

Instructions: Have everyone get into their groups and work through the challenges of their assigned roles. The anthropologists will wait and discuss their "study" of this topic and then go around from group to group taking notes on what everyone thinks about their situation. Then they report back to the group at-large their findings, so all the groups benefit from the larger scenario.

Critical thinking question to consider: How did you feel about the assigned roles? How do you feel about having the roles of the American bosses and the Mexican bosses being assigned to males? Does this reflect the reality of factory work in general?

Field Notes

INDIVIDUAL ACTIVITY: IT'S THE THOUGHT THAT COUNTS

©iStockphoto.com/HAKINMHAN

Note: This can be a reflection assignment that students write and turn in or a forum quiz/post in their Learning Management System.

Read: Reciprocity is the idea that there is intrinsically a relationship and expectation when gifts are given. The simplest way to see this is to think about when you give a good friend a birthday gift. Then when your birthday rolls around, you wait for them to give you something. If they don't, would you be disappointed? What would that tell you about your relationship or about that friend?

Reciprocal relationships do not have be equal in terms of money or even sentiment of the gifts that are given. Think about a child getting a bicycle from their parents and then giving their parent a macaroni necklace in return. The necklace is not equal in price to the bicycle, and nor is the bicycle equal in terms of sentiment to the necklace, yet they can be equivalent gifts. You don't see parents on Christmas morning thinking that their 7-year-old child should really get his or her act together to give them better gifts. Reciprocity demonstrates relationships between the gift givers. Gifts are the tangible evidence of the social relationships between givers and recipients. They often help us understand other people and our relationships with them.

Activity: Think about your own gift giving in the past.

Question 1: Who do you give gifts to and on what kinds of occasions?

____

____

____

Question 2: When would your feelings be hurt by not receiving something in return to match your gift?

____

____

____

Question 3: When would you accept the imbalance in the gift giving?

____

____

____

FIELD WORK

This week, find someone in your life who you would not normally have a "gift-giving" or close relationship with and give them a small gift.

Field Notes

What was their reaction?

Do you expect that they will counter this gift with something in return at some point, or will they just receive your gift and move on?

Were they surprised by your gift? How did it make you feel?

EXERCISE SEVEN

KINSHIP

Key Terms

Affinal kin: ______________________________

____

Consanguineal kin: ______________________________

____

Fictive kin: ______________________________

____

Kinship: ______________________________

____

Kinship terminology: ______________________________

____

CLASSROOM ACTIVITY: FAMILIES ARE COMPLICATED

Note: This activity can be assigned as a classroom discussion in a small class or in groups in a larger class. It may also be administered by assigning students to complete and reflect.

Read: Anthropologists use kinship charts to find out information about other cultures' families and relationships. The average American family keeps track of about five generations (from their grandparents to grandchildren). The Nuer in Sudan kept track of up to 11 generations of family, and that might seem like an impossible task to us without the help of an ancestry research company, genealogy archives, or talking to your family and extended family. Kinship charts show us family relationships, and especially who is related by blood or marriage when those aren't always clear to an outsider.

Families are complicated. People may have babies with several different partners and have relationships that are not always official in any way. They may have whole extended families that they do not know or relatives no one speaks of anymore. Anthropologists know that there is so much history tied up in family stories and also culture in the patterns that families create. Some cultures dictate that mothers' sisters also be called "mother," and then they serve as maternal figures for children, to be respected the same as their biological mothers. Imagine you walk into a culture and you learn that a child has five mothers, and they think you are the crazy one for not understanding their system. This is why kinship charts are so important to us.

Activity: Making your own kinship chart.

Instructions: When you are creating the chart, go up to your grandparents and down to your children or your siblings' and cousins' children, one generation below yours. Spend about 10 minutes working out your family, and then pair up with a classmate and explain your families to each other to check your work. The chart should make it clear to your classmate who everyone is in your family if you have drawn it correctly. Fix any mistakes you find.

Here is an example of a completed kinship diagram:

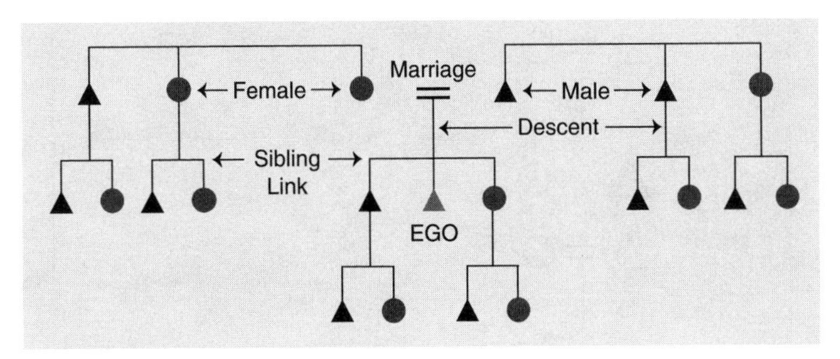

This chart is a consanguineal chart, so it only shows the blood relatives of EGO. Make sure to include your affinal kin who have married into your family, like the uncle who married your mother's sister and is the father to you cousins.

Here is an example of the kinship key:

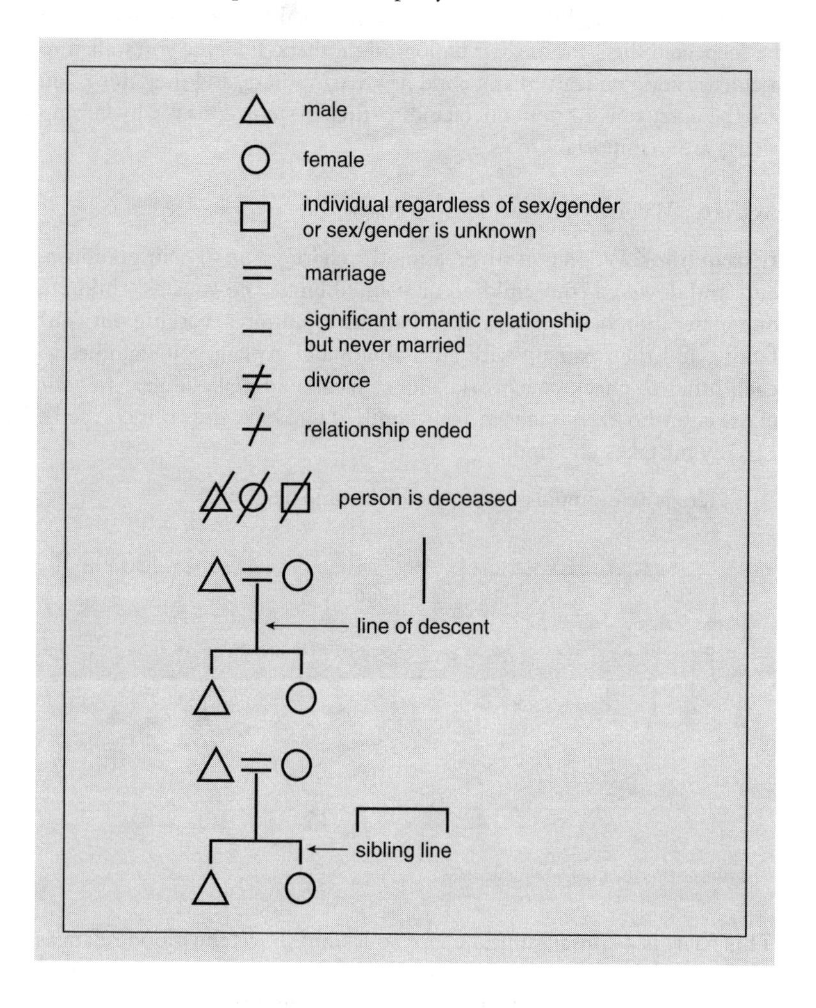

Getting started: Ego is the person whose chart you are creating. In this case, you are Ego.

For example, if you are female: ○

Ego

Field Notes: Kinship Diagram

INDIVIDUAL ACTIVITY: CHOOSING TO BE FAMILY

©iStockphoto.com/Wavebreakmedia

Note: This can be a reflection assignment students write and turn in or a forum quiz/post in their Learning Management System.

Read: Kinship is often traced through descent or marriage, but what about other people whom you consider to be family but who do not fit that criteria? Anthropologists call those relationships fictive kin. We assign them family relationships and monikers even though these people aren't related to us by descent or marriage. In the United States, many children grow up with adults they refer to as uncle or aunt but who are not siblings of either their mother or father (the conventional meaning of these words in US society). Maybe they are your father's or uncle's best friend, and they have just become part of your family. Maybe they are your godparents, friends whom your parents decided to make part of your family through this relationship. These relationships are often important ones in our lives, and they can be just as important, if not more important figures in your family, than someone who was born into it.

Question 1: Who are the fictive kin in your life?

Question 2: How did you find out that you have this kind of relationship?

Question 3: Is this person or are these people meaningful to you in your life? If so, how?

Read: People in the United States often consider their pets to be part of the family. Now, hopefully these pets are not related to you by blood or marriage (ha!), so how would you define that relationship? Usually those kinship relationships often still fit into the role of fictive kin because people talk about their "fur babies" or "grand puppies" as if they are some form of children or grandchildren.

Question 4: Does this apply to your pets?

____

____

____

____

____

Question 5: What kinds of pets do you have or have you had?

____

____

____

____

____

Question 6: Why do you consider your pets to be family?

____

____

____

____

____

Question 7: Would the kinship relationship you have with your pets change based on the type of pet that it is? Does a horse or snake or chicken become any more or less family than a dog or cat? Are fish able to be family members? What about gerbils or birds? A skunk? A turtle? Is there a limit to what you would count as family?

EXERCISE EIGHT

MARRIAGE

Key Terms

Endogamy: ______________________________

Exogamy: ______________________________

Marriage: ______________________________

CLASSROOM ACTIVITY: PICKING A PARTNER

©iStockphoto.com/AygulSarvarova

Note: This activity can be assigned as a classroom discussion in a small class or in groups in a larger class. It may also be administered by assigning students to complete and reflect.

Read: Anthropologists love to study marriage and the rules around marriage. There are many different types of marriage patterns, including one where after the marriage the wife moves in with the husband's family (virilocal), one where the husband moves in with the wife's family (uxorilocal or matrilocal), and even one where no one leave their natal home after they marry (natalocal). Imagine marrying someone and never leaving your parents' home, ever. Marriage rules around whom you can marry can also be complicated. *Endogamy* and *exogamy* are terms that denote marriage rules for whom you can marry within your culture. A group that practices endogamy supports marriages only among its own members. A group that practices exogamy supports marriages only with members of other groups. Groups included in rules of endogamy and exogamy can include cultural groups, ethnic groups, religious groups, groups defined by educational levels, extended family, and many other qualifiers. The idea of endogamy helps keep partners within the same cultural group, encourages them to have the same cultural practices or religion, and helps to ensure that people continue to stay within a geographic boundary, not moving or marrying out of it. The idea of exogamy encourages new genetic material, preventing inbreeding in royal families, and it can create new relationships across other towns and neighboring villages. If a family member from one village marries into another, they can help to smooth out tensions or arguments between the groups and act as a bridge between them. Marriages are partnerships not only between people but between families and sometimes entire cultures.

Activity: Think about your marriage, current relationship, or past and future partners.

Were or are there expectations that you would date a member of particular group of which you are also a member? Did these expectations come from you or your family? Do you think about this differently than your parents?

Discuss this in groups of four or five students and then together as a class. Are there similarities across the groups or the entire class?

Field Notes

____

____

____

____

____

____

INDIVIDUAL ACTIVITY: WHAT'S THE RUSH?

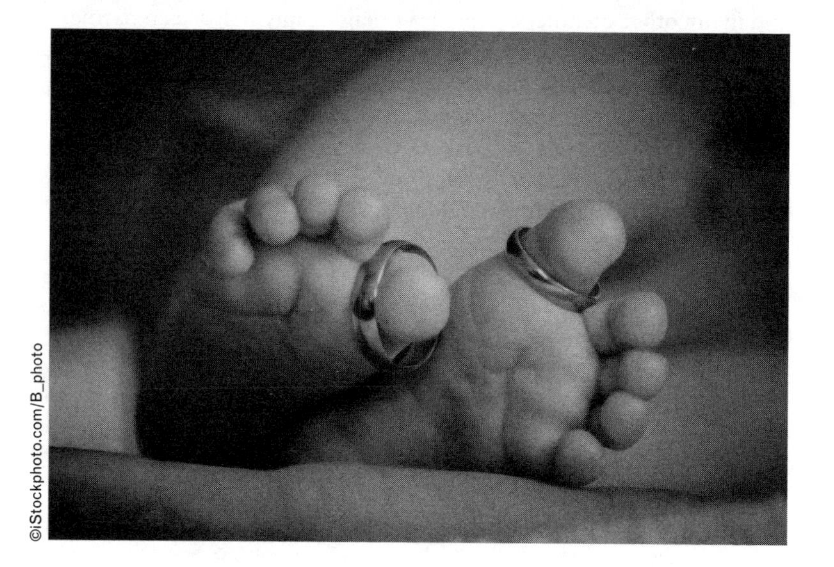

Note: This can be a reflection assignment that students write and turn in or a forum quiz/post in their Learning Management System.

Read: There has been a rise in delayed marriage and delayed child bearing in the United States over the last 40 years. The US Census reports that between 1947 and 1975, the median age at which Americans got married for the first time ranged between 20.5 and 21.5. However, since then, it has risen steadily. By 2000, it had risen to 25 and by 2015 to more than 27. Furthermore, the Pew Research Center's study on marriage shows that in 1970, nine in ten 30-year-olds had married, but by 2008 the number is six in 10. The average age of first-time mothers, married or not, has also steadily increased.

Americans are less interested in early-age marriage, while other parts of the world, like India and Yemen, still engage in the practice of taking child brides. Many different factors make this a complicated picture. Socioeconomic factors, gender role expectations, the large cost of weddings, cultural and historical practices, fertility and childbearing expectations, and legal statutes forbidding or outlawing certain practices all add to the differences in marrying and having children.

Question 1: Why do you think the age at which people marry has been delayed in the United States?

____

____

____

Question 2: How do you feel about getting married?

____

____

____

____

____

Question 3: What is the ideal age to get married?

____

Question 4: What is the ideal age to have children?

____

Question 5: How many children do you want to have?

____

Question 6: What factors would affect the number of children you would like to have? (e.g., finances, partner, etc.)

____

____

Question 7: Why do you think this is different in other countries and cultures?

Question 8: What might affect this across the world?

EXERCISE NINE

GENDER

Key Terms

Gender: ______________________________

____

Sex: ______________________________

____

Sexuality: ______________________________

____

LGBTQIA: ______________________________

____

Trans*: ______________________________

____

Two-spirit: ______________________________

____

Hijra: ______________________________

____

CLASSROOM ACTIVITY: A RAINBOW OF POSSIBILITIES

©iStockphoto.com/DanielBendjy

Note: This activity can be assigned as a classroom discussion in a small class or in groups in a larger class. It may also be administered by assigning students to complete and reflect.

Read: In American culture, we often think of gender as a binary between male and female, so that even when people identify as transgender, there is an assumption they will demonstrate some connection to masculinity, femininity, or a mixture of both. However, around the world, there are many cultural gender and sex roles like the hijra of India and the two-spirits of Native North American tribes that have created culturally appropriate roles that allow people to identify as a third sex or gender. These roles are historical, cultural, and considered perfectly normal and so culturally integrated that despite the attempts of colonial and modern governments to abolish them, they have persisted.

In the United States, the LGBTQIA and especially the Trans* community are gaining larger and wider acceptance in recent years. In 2014, Facebook added more than 50 terms for users to define themselves with regard to their sex gender ideologies and even privacy protections, so users can regulate who sees them. Two-spirit is one of those terms! Here is the full list:

- Agender
- Androgyne
- Androgynous
- Bigender
- Cis
- Cis Female
- Cis Male
- Cis Man
- Cis Woman
- Cisgender
- Cisgender Female
- Cisgender Male
- Cisgender Man
- Cisgender Woman
- Female to Male
- FTM
- Gender Fluid
- Gender Nonconforming
- Gender Questioning
- Gender Variant
- Genderqueer
- Intersex
- Male to Female
- MTF
- Neither
- Neutrois
- Non-binary
- Other
- Pangender
- Trans
- Trans*
- Trans Female
- Trans* Female
- Trans Male
- Trans* Male
- Trans Man
- Trans* Man
- Trans Person
- Trans* Person
- Trans Woman
- Trans* Woman
- Transfeminine
- Transgender
- Transgender Female
- Transgender Male
- Transgender Man
- Transgender Person
- Transgender Woman
- Transmasculine
- Transsexual
- Transsexual Female
- Transsexual Male
- Transsexual Man
- Transsexual Person
- Transsexual Woman
- Two-Spirit

Activity: Get into groups to see how well you know all of these terms and what they mean. Look up any terms you don't know. Our terminology is always changing (remember the Sapir–Whorf hypothesis!), and we create words for things we need to name.

Discuss as a group whether there are any terms missing from Facebook's list. What terms are gaining or losing popularity? What are the benefits for someone to be able to identify as their specific gender identity in public or private on Facebook? Are there possible costs as well? Discuss as a class.

Field Notes

INDIVIDUAL ACTIVITY: A VERY PERSONAL INTERVIEW

©iStockphoto.com/Steve Debenport

Note: This can be a reflection assignment students write and turn in or a forum quiz/post in their Learning Management System.

Read: Gender, sex, and sexuality are often privately defined concepts of self, which are then publicly presented on our bodies. Keep in mind that these three ideas of gender, sex, and sexuality are all different concepts, and one does not hinge on the other. Gender refers to the cultural, social, and psychological definition of self. Sex refers to the biological, reproductive, hormonal, and genetic parts and pieces one has on and in their body. Sexuality pertains to whom you want to have sex with and what kind of sex that might be, or lack thereof. How you choose to define yourself within these three ideas can change over time. Use this space and opportunity to get to know yourself better. Consider your answers carefully and write them in the space below.

Question 1: How do you define your own sex, gender, and sexuality?

Question 2: How did you know who you are, and what terms you would like to use for yourself?

Question 3: Have you faced any obstacles in your journey to understanding your own sex and gender? Please explain.

Question 4: Do you dress in a way that is intentionally masculine, feminine, mixed, or neutral?

FIELD WORK

Now, with these same questions, interview someone in your life, maybe a friend or family member, who identifies based on gender or sex or sexuality differently than you do. Make sure to interview someone who will be comfortable speaking with you on this topic. Ask them each question and put their answers below. Remember, interviews are always a conversation, so feel free to deviate and speak about other related examples and topics, but come back to the questions below to ensure everything is fully addressed by the interviewee.

Field Notes

Questions for interviewee:

What is your first name?

____

How do we know each other?

____

____

How do you define your own sex, gender, and sexuality?

____

____

____

How did you know who you are, and what terms you would like to use for yourself?

____

____

Have you faced any obstacles in your journey to understand your own sex and gender?

Do you dress in a way that is intentionally masculine, feminine, mixed, or neutral?

Summarize what you learned about this person in your life from this interview. Did their answers surprise you? Did you feel comfortable interviewing this person?

EXERCISE TEN

POLITICS

Key Terms

Band: ____________________

____

____

Tribe: ____________________

____

____

Chiefdom: ____________________

____

____

State: ____________________

____

____

CLASSROOM ACTIVITY: FIND ONE, FIND ALL

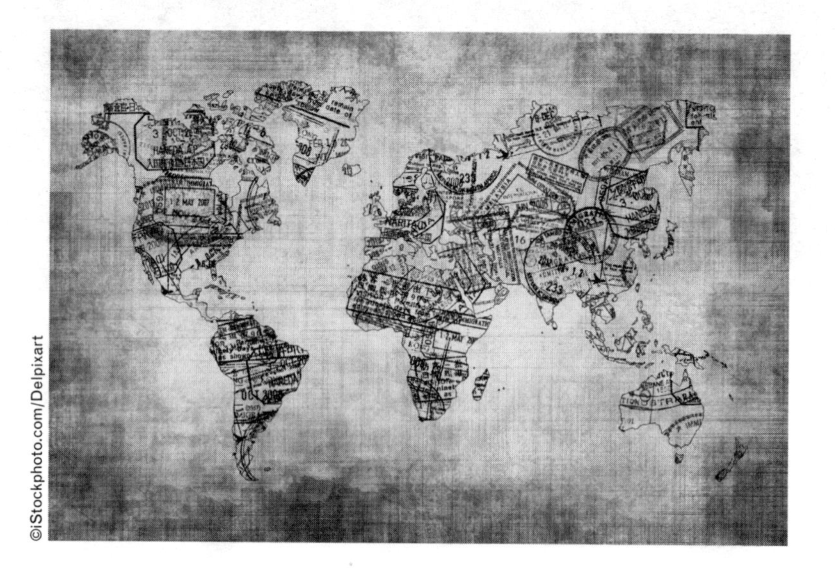

Note: This activity can be assigned as a classroom discussion in a small class or in groups in a larger class. It may also be administered by assigning students to complete and reflect.

Read: Bands, tribes, chiefdoms, and states all function differently as political structures for society. Some of that has to do with scale, as bands are going to be smaller than states in size and population, but they also differ in terms of needs of societies to keep these organized populations successful. States require a lot of infrastructure and hierarchy to keep them functioning. In each of differing parts of societal organization, there are levels of power and authority necessary to enforce and maintain the structure. Chiefdoms have to have a chief, whereas bands are often made up of a family, and the band is led by either a mother or father figurehead, depending on whether it is matriarchal or patriarchal. States consolidate their power and authority into a single leader, but they also have a large political organization that supports the leader and works to maintain their power over the people they represent.

Activity: Divide into groups of four.

Each group should be assigned a different country from Latin America or Asia. As a group, use your web-searching ability to find one band society, one tribe society, one chiefdom, and one state all in that single country.

Groups should present findings to the class or post in a Learning Management Strategy (LMS) forum with their web links; then students can comment on other groups' findings.

Please consider and answer these questions:

Who are these cultures? Why do they fit the criteria for each stratified level of organization? If you can't find one, why not?

Field Notes

INDIVIDUAL ACTIVITY: CHOOSE YOUR OWN UTOPIA

©iStockphoto.com/FilippoBacci

Note: This can be a reflection assignment students write and turn in or a forum quiz/post in their LMS.

Read: Societies become formed in the images they are culturally grounded within, meaning family-based groups that need land and space for cultivating crops and raising animals cannot be too close in proximity to others because of their physical needs and often stay on the smaller scale, usually as tribes or chiefdoms, maybe never becoming states. States require things like labor forces and divisions of tasks so that some people can spend all their time in charge of other people, not working their land. There are many ways in which each political and social organization strategy is unique depending on the type of structure within a particular society.

Activity: Imagine that there has been an apocalypse, and you run off into the woods with some friends and family. From there, you can design your perfect world as you begin to start over. Is it a small family-based band society, or do you want to immediately install yourself as the head of your own chiefdom? Think about your options and write which society organization you have chosen and why you have not selected the others.

Question 1: What kind utopia are you imagining? Where are you? What does it look like? Who is there with you?

____

____

____

____

____

Question 2: Who is in charge of your utopia? Is there a leader?

____

____

____

Question 3: How do you get food to survive? What is your subsistence strategy?

____

____

____

____

____

____

____

Select one of the following to describe the organization of your utopia, and for each of the others, explain why you have not chosen it.

Band

Tribe

Chiefdom

State

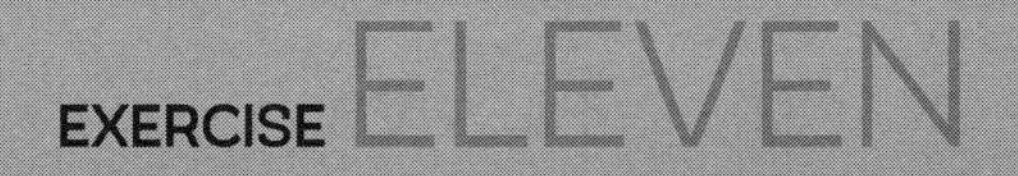

EXERCISE ELEVEN

STRATIFICATION

Key Terms

Class: ____

____

Class system: ____

____

Social mobility: ____

____

Social stratification: ____

____

Wealth: ____

____

CLASSROOM ACTIVITY: MONEY CAN'T BUY YOU CLASS, OR CAN IT?

Note: This activity can be a reflection assignment students write and turn in or a forum quiz/post in their Learning Management System (LMS).

Read: Income is a big factor in class and class creation around the world, but especially in the United States. When we engage in discussions of class, often we only think about it terms of wealth, as in low, middle, and upper class by their financial means. But socioeconomic factors go together, hand in hand. What does that actually mean for people? What social factors also combine with economics to make it a more complicated picture? These can be things like race, sex, religion, marital status, number of children, education, and many others. These social factors make a lot of difference in establishing your place within an economic structure and also your ability to move within the economic structures and from class to class. They can allow, encourage, inhibit, and actively prevent you from moving up the ladder. Think about how hard it would be to be a wealthy African American woman during segregation in the southern United States in the

1950s and what factors would inhibit her ability to move up socially and economically. Would some of these factors still apply to that same African American woman as a college student today? Some social factors change faster over time as society and culture change, and some more slowly.

Activity: In small groups, discuss and answer these questions:

- Can people in the United States easily move up and down in class (from lower to middle class or from middle to upper class, for example)?
- What factors might affect this?
- What is the relationship between education and class?
- What is the relationship between race, ethnicity, and class?
- In what ways does the class into which you are born affect your ability to move up? What factors make it harder to move out of poverty and into the middle or upper class?
- What factors make it easy for people born into the upper class to remain there?

Then as a class, come together to discuss the factors that affect class. Share some of the personal examples that came up in the smaller group discussions. If it is a large class, have the groups write down their lists and hand them in for later review.

Field Notes

INDIVIDUAL ACTIVITY: WHAT IF?

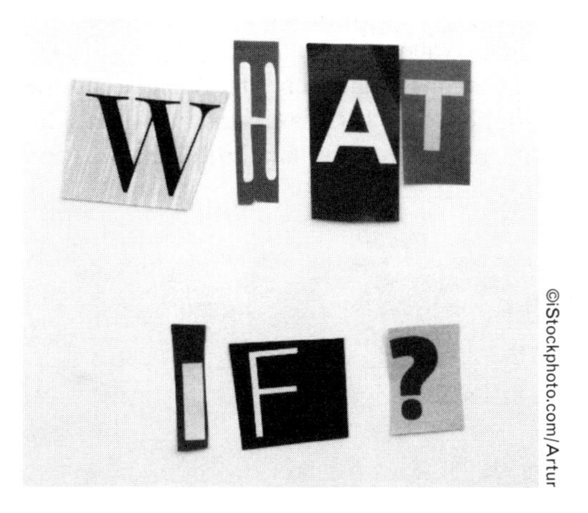

©iStockphoto.com/Artur

Note: This can be a reflection assignment students write and turn in or a forum quiz/post in their LMS.

Read: When we think about class, race, sex, gender, economics, politics, religion, and other identifiers in our culture, we have a variety of ways to define ourselves and to be defined by others. Most of these are immovable, unchangeable characteristics we carry around, but some like class and economics can change if we work hard and other circumstances allow. As discussed in the previous exercise, these factors combine to make our own personal situations, and often some will cause restrictions, preventing access to economic and social mobility. In your own life, have you thought about how much easier life would be *if*... (insert thing here)? What is that *if*? Is it money? Is it power? Is it education? Is it religion, race, or ethnicity? Is it being from somewhere else?

If, in all seriousness, you had the chance to change this thing about yourself, would you do it? What sorts of things would you be giving up if you did?

FIELD WORK

Activity: Interview your parent, grandparent, or another family member who is in a generation above you.

Ask them questions about their "if." What would they change that they think might help change their class status? Create five questions to ask them specific to you and your family. Put the questions and answers below.

Field Notes

Interviewee first name:

Relationship to you:

1. Question ______________________________

Answer:

____

____

____

____

2. Question ______________________________

Answer:

____

____

____

____

3. Question ______________________________

Answer:

____

____

____

____

4. Question ______________________________

Answer:

____

____

____

____

5. Question ______________________________

Answer:

____

____

____

____

EXERCISE TWELVE

RELIGION

Key Terms

Magic: ______________________________

Shamanism: ______________________________

Religion: ______________________________

Ritual: ______________________________

Sacrifice: ______________________________

Sorcery: ______________________________

Totem: ______________________________

Witchcraft: ______________________________

____

____

CLASSROOM ACTIVITY: MY LUCKY SOCK!

©iStockphoto.com/LightFieldStudios

> **Note:** This activity can be assigned as a classroom discussion in a small class or in groups in a larger class. It may also be administered by assigning students to complete and reflect.

Read: Magic, shamanism, sorcery, and rituals may seem like strange, far away practices that other people do. But if you are studying for an exam, and you *always* do x, y, and z, then that is a ritual you do for luck on the exam. Most people will be familiar with our superstitions in American culture, things like avoiding walking under a ladder, not breaking a mirror, or not letting a black cat cross your path. We also have many rituals and magic practices that we do, especially around things like sports. From Little League to professional sports, from volleyball to track and field to football and fútbol, we see lots of magic, religion, rituals, chants, and so on.

Activity: Get into groups. The instructor will assign each group a different sport.

Use the Internet to look up the famous superstitions in each sport. What kinds of rituals are the athletes doing to help their game? Are there songs, food, words, chants, magic potions, special socks, and the like, that have been used, loved, or banned over the years by players or teams? Look at both famous athletes' individual lucky charms and also team charms, practices, and rituals.

Make a list of the things you find in groups, and then present to the class your examples. These can also be posted to a group forum in a Learning Management Strategy (LMS) with discussion and commentary by other groups.

Field Notes

INDIVIDUAL ACTIVITY: YOU GOTTA HAVE FAITH

Note: This can be a reflection assignment students write and turn in or a forum quiz/post in their LMS.

Read: Many people are raised in homes in which religion is an important practice. Some may question the beliefs they are taught as children, while others never do. You may have been raised in a specific belief system or religion and frequently worshipped at a church, synagogue, mosque, or temple. Or, you may have never set foot in one at all. About 85% of the world's population identifies with a religion. So it's safe to say that religion is at least somewhat important to the vast majority of people.

FIELD WORK

Activity: For this activity, you need to visit a house of worship or religious center and attend a service.

If you grew up identifying as a member of a specific religious group, step outside your own box and go to the worship service of a different religion or a different denomination of the religion with which you identify. Make sure to do some research before you go so that can observe their customs. They may ask you to remove your shoes, so you might need to ensure you have socks on. Be aware that many places have regulations concerning dress. Women may be required to wear clothing that covers their shoulders and knees. Men may not be permitted to wear shorts, t-shirts, or other informal attire. Check before you go. Ask lots of questions so you can be as respectful as possible to someone else's place of worship.

While there, look around. Observe and make notes.

Field Notes

Who is attending these services? What kind of message is being delivered? Are there songs? Is there literature being read from a holy book? Who gets to speak or sing or participate?

____

____

____

____

____

____

____

____

How is the space arranged? What does the room look like? Sketch out a map of the space where the worship service is being held. Make sure to mark where you are.

After your visit, summarize the experience here below. How did you feel? How similar or dissimilar was this to your own beliefs? Will you go back?

EXERCISE THIRTEEN

ART

Key Terms

Arts: ____________________

____

____

____

Deep play: ____________________

____

____

____

World art: ____________________

____

____

____

CLASSROOM ACTIVITY: THE ART OF THE RIVALRY

Note: This activity can be assigned as a classroom discussion in a small class or in groups in a larger class. It may also be administered by assigning students to complete and reflect.

Read: Deep play is the idea that participants and spectators of an event are joined together in an important cultural process. Deep play events manipulate important cultural themes such as social stratification, masculinity and femininity, power and authority, and processes of production. Often, these are sporting events or cultural events. Some examples are sumo wrestling in Japan, cock fighting in Bali, or cricket in Britain (and in places like India that were formerly part of the British Empire).

What are some American cultural events or artistic endeavors that could be considered example of deep play? Some signs that an event may be deep play are that the event attracts a relatively large number of people,

people care passionately about the outcome of the event, the event is in some way connected to critical cultural and social issues (often this connection is symbolic or metaphoric), and that ultimately, the event does not change society.

Football is a classic American example of deep play. It speaks to the violence and sexuality underlying competition between men, the social role of women, the relationship of the individual to the group, structured labor, rules and their infringement, gaining and surrendering territory, and racial character. However, not all football games are equally deep or important. Some involve great passions among fans, others not so much. In many cases, in college football, there are famous rivalries. Some of the greatest college football rivalries include Army/Navy, Alabama/Auburn, Michigan/Ohio State, and Texas/Oklahoma. Games between these teams seem far deeper than most other games.

Activity: Get into groups. Consider, does your university have a sports rivalry (if you go to a smaller school and are not a sports fan, you may need to research this)? What sport is it in?

____

____

Are games played against your rival school deep? Why or why not? What is really being communicated in these games?

____

____

____

____

____

If your school does not have a rivalry like this, consider a professional sports league rivalry for this exercise. Do you know of or participate in a rivalry for the National Football League, National Basketball

Association, National Hockey League, Major League Baseball, or Major League Soccer?

INDIVIDUAL ACTIVITY: MAKE YOUR MARK

©iStockphoto.com/absolutimages

> **Note:** This can be a reflection assignment students write and turn in or a forum quiz/post in their Learning Management Strategy.

Read: Art, and what is considered art, changes based on where you are in the world and who you are. Children make a lot of "art"! Historically, in Western culture, only painting, sculpture, architecture, music, and poetry were included in the fine arts. Only recently has the idea of art grown to include a wide variety of things. Now, it is widely accepted in American culture that art can be carved, burned, and inserted into and on your skin with tattoos, piercings, and branding. Tattoos especially have long been identified with everything from military service to prison, gangs, and criminals, and over time they became a form of artistic expression for both those who tattoo and those who get tattoos. There is even an art form that requires immense physical skill and mental energy called suspension where people use strategically placed piercing holes in their bodies along with hooks, straps, and wires to suspend themselves from in private spaces and also in public exhibitions. Suspension is becoming more mainstream art, having left performance artists and moved to the general public. Look at the image below and if you are intrigued, search images for suspension online, but avert your eyes if they are sensitive. It can look pretty terrifying and painful.

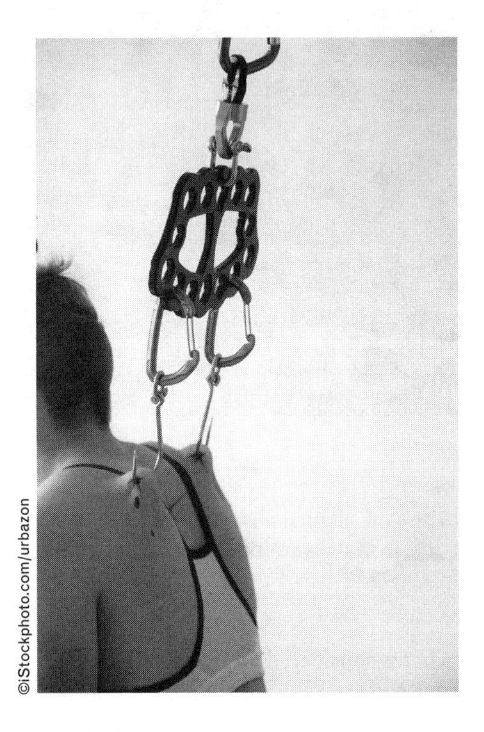

Activity: Tattoos, piercing, branding, and suspension are seen as mainstream forms of artistic display. Much of this skin-deep artistic expression no longer generally disqualifies you from a career in teaching children or working at a bank, as they would have even 10 years ago. There are some exceptions, such as hand and face tattoos and excessive piercings or gauged ears, that may still be problematic for someone in America today.

Question 1: Do you have tattoos or piercings?

Question 2: What do you think changed the view on tattoos and piercings in American culture?

Question 3: Do you think of these as forms of art?

Question 4: Do these add to, take away from, or help create identity for people who have them?

Question 5: Tattoos and other forms of body modification are common around the world. Why do you think this is so?

EXERCISE FOURTEEN

POWER

Key Terms

Colonialism: ______________________________

____

____

____

Colony: ______________________________

____

____

____

Power: ______________________________

____

____

____

CLASSROOM ACTIVITY: DISCOVER COLONIALISM

Note: This activity can be assigned as a classroom discussion in a small class or in groups in a larger class. It may also be administered by assigning students to complete and reflect.

Read: Much of the world was affected by European expansion and colonialization. In many cases, people continue to live with its consequences. The effects still exist today where diseases continue to circulate, where people continue to work in indentured servitude, where non-native crops are cultivated, and where indigenous people live in subservient roles to colonizing populations, even still left out of their own governance. The photo above is of the Elmina Slave Castle in Ghana in West Africa. The door is the entrance male slaves passed through to holding cells that had little ventilation and no bathroom facilities. Men were held there for days on end until being packed into slave ships and sent to the New World on the Atlantic Slave Trade. Elmina and the nearby Cape Coast slave castles are both now United Nations Educational, Scientific and Cultural Organization World Heritage sites with hundreds of thousands of visitors each year, who come to remember what happened and the long-lasting consequences of colonialism. The same could be said for thousands of sites for native peoples around the globe, where atrocities have occurred in the name of colonialism. The recent American advocacy to change the name of the, federal holiday Columbus Day to Indigenous Peoples Day shows some raised awareness for the lasting role of colonialism here in America for Native Americans and First Nations.

Activity: Get into small groups. Every groups should select (or be assigned) a different country across Africa and Latin America. Then look online to answer the questions below about your country.

Share your results in class presentation or in a forum on a Learning Management Strategy (LMS) with discussion and comments.

Field Notes

Where is your country, and who are its modern neighbors?

____

____

Who (it may be more than one) colonized this country and its people (or tried to)?

____

____

When did it gain its independence?

____

What specific consequences did this country face during colonialism?

What specific consequences did this country face after colonialism?

Are there any modern repercussions still affecting it today?

INDIVIDUAL ACTIVITY: IMAGINE THE OUTCOME

©iStockphoto.com/SL_Photography

Note: This can be a reflection assignment that students write and turn in or a forum quiz/post in their LMS.

Read: As discussed in the previous class assignment, the effects of colonialism are still felt today by indigenous people across much of the world. But often race and culture play a large part in determining whether the outcomes of colonialism affect people in a positive or negative light. Many postcolonial countries still face massive levels of national poverty and are in debt to financial organizations like the World Bank or the International Monetary Fund. They may still be plagued with disease, struggle with natural resource procurement, and have cultural or ethnic conflict, usually a colonial remnant from how countries were carved up upon the departure of colonial governments.

Activity: Take the country from the classroom exercise that you have already researched and think through two scenarios. If your class skipped the above exercise, then add that to this task and then continue below.

In the first scenario, you are a part of the descended population from the colonizer of this country. How might things be for you in said place as this person if you were born there now?

In the second scenario, you are a part of the colonized local cultures that were deeply affected by colonial rule. How might things be for you in said place as this person if you were born there now?

Are these two scenarios different for these two imaginary versions of you? How does this make you feel?

EXERCISE FIFTEEN

CULTURE CHANGE

Key Terms

Development: ____________________

____

Export processing zones: ____________________

____

Global assembly line: ____________________

____

Multinational corporation: ____________________

____

Sweatshop: ____________________

____

CLASSROOM ACTIVITY: WHO ARE YOU WEARING?

©iStockphoto.com/Tero Vesalainen

Note: This activity can be assigned as a classroom discussion in a small class or in groups in a larger class. It may also be administered by assigning students to complete and reflect.

Read: The goods and products we buy now come from all around the world. We have already explored what happens in a maquiladora, but what about other export processing zones? The vast majority of consumer products that we buy, everything from televisions to t-shirts, is manufactured outside the United States, usually in less wealthy nations. Anthropologists have studied the production of raw materials, processing, manufacturing, assembly, transportation, and final points of sale for many goods. Because the processes that create the products we buy often take place in many different countries, we often refer to the whole process as the global assembly line. Multinational corporations are involved at every stage of this process, and large transportation networks move goods around the globe as they transform from raw materials into useable consumer goods.

Activity: Everyone look at the tag of the shirt you are wearing and find out where it was made. Form groups based on the country of your shirt's production. Then in those groups, research the conditions for clothing workers in your country. Then look for information on the conditions for workers who make the clothing for the brands you are wearing. Some clothing stores or brands will have more information readily available online than others.

Field Notes

Get the groups together who are in neighboring countries. Are the conditions and issues for workers similar across this region? Why or why not?

Get the continents together to discuss the similarities. Are the conditions and issues for workers similar across the continent? Why or why not?

Then discuss as a class.

INDIVIDUAL ACTIVITY: WHAT MATTERS MOST?

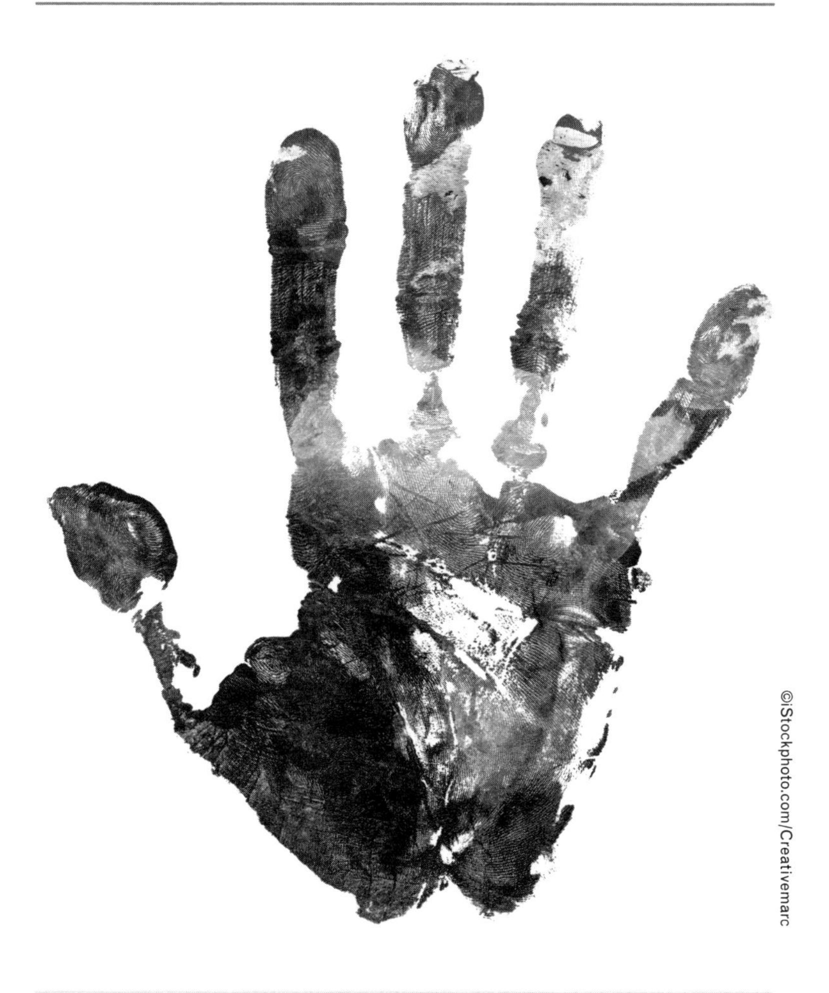

©iStockphoto.com/Creativemarc

Note: This can be a reflection assignment that students write and turn in or a forum quiz/post in their Learning Management Strategy.

Read: There are many social, environmental, and political issues happening around our world today that ask for our attention. From immigration to climate change, from child labor to indigenous issues, there are many challenges we face that all need studying. Anthropologists often choose narrow fields of study, and that means there are things that will always need more research and advocacy.

Activity: If you were an anthropologist working today, what would be your priority for a study on a modern challenge? What issues are most important to you?

What challenges would you face in trying to study this?

Where in the world would it take you?

Why is this issue so pressing?

SUMMARY EXERCISE

INDIVIDUAL ACTIVITY: WHAT MAKES A GOOD ANTHROPOLOGIST?

©iStockphoto.com/seb_ra

Note: This activity can be a reflection assignment that students write and turn in or a forum quiz/post in their Learning Management Strategy.

Read: By now, you have seen lots of examples of what topics anthropologists study and how anthropology has been involved in many countries and cultures. In the very beginning of the class, you learned that anthropologists always try to remain culturally relative and refrain from any judgement or ethnocentrism. But there are many other things anthropologists need to do to address the needs and issues of the cultures they study.

Activity: List below the traits and characteristics of a good anthropologist.

Why is anthropology an important discipline?

Would you want to be an anthropologist? Why or why not?

The Field Journal for Cultural Anthropology

For H. with love.